ELECTRIC ANIMALS

PLATYPUSES SENSE ELECTRICITY!

BY LOUIS MALLORY

Gareth Stevens
PUBLISHING

Please visit our website, www.garethstevens.com. For a free color catalog of all our high-quality books, call toll free 1-800-542-2595 or fax 1-877-542-2596.

Cataloging-in-Publication Data
Names: Mallory, Louis.
Title: Platypuses sense electricity! / Louis Mallory.
Description: New York : Gareth Stevens Publishing, 2024. | Series: Electric animals | Includes glossary and index.
Identifiers: ISBN 9781538292990 (pbk.) | ISBN 9781538293003 (library bound) | ISBN 9781538293010 (ebook)
Subjects: LCSH: Platypus–Juvenile literature.
Classification: LCC QL737.M72 M355 2024 | DDC 599.2'9–dc23

Published in 2024 by
Gareth Stevens Publishing
2544 Clinton Street
Buffalo, NY 14224

Designer: Claire Wrazin
Editor: Natalie Humphrey

Photo credits: Cover, pp. 1, 7 Lukas_Vejrik/Shutterstock.com; background (series art) Romashka2/Shutterstock.com; p. 5 worldswildlifewonders/Shutterstock.com; p. 9 (platypus) Tammy27/Shutterstock.com, (echidna) Mari_May/Shutterstock.com; p. 11 bluesmoke/Shutterstock.com; p. 13 Susan Flashman/Shutterstock.com; p. 15 trabantos/Shutterstock.com; p. 17 John Carnemolla/Shutterstock.com; p. 19 OBSESHNZ/Shutterstock.com; p. 21 Dave Watts/Alamy Stock Photo.

Printed in the United States of America

CPSIA compliance information: Batch #CW24GS: For further information contact Gareth Stevens, New York, New York at 1-800-542-2595.

CONTENTS

Boldface words appear in the glossary.

A Strange Mammal

The platypus is a strange-looking **mammal**! It's usually found in freshwater rivers and lakes in Australia. But the platypus doesn't just look strange, it also has a strange superpower. Platypuses can sense **electricity**!

The Platypus Body

Platypuses have short, brown bodies covered in **coarse** fur. Their feet are wide and **webbed** to help them move through the water. Platypuses can be up to 25 inches (63 cm) long and weigh up to 6.6 pounds (3 kg).

Monotremes

Platypuses are part of a group of animals called monotremes. Monotremes are the only mammals that lay eggs. Monotremes are also the only known mammals that can sense electricity. Echidnas are also part of this family.

ECHIDNA

The Bill

A platypus has a soft bill that looks like a duck's bill! A platypuses' bill is covered with **receptors**. Their receptors are used to sense electricity. A platypus can have over 40,000 receptors in their bills! These receptors work best underwater.

Sensing Electricity

How do these receptors work? When an animal is moving, its **muscles** make a small amount of electricity. The receptors in the platypuses' bill can sense this electricity. The receptors help the platypus find out how far away their food is.

On the Hunt

When it's time to hunt, platypuses close their eyes and ears and dive underwater. They can hold their breath for up to two minutes when they hunt. They move their head back and forth in the mud on the bottom of the river or lake until their receptors sense food. Then, they strike!

Bugs for Dinner

Platypuses mostly eat bugs that have a softer body. They use their bills to dig bugs out of the ground underwater. Platypuses keep these bugs in **pouches** in their cheeks until they are ready to come back to the **surface**.

Toothless

Platypuses don't have any teeth. Instead, they have a hard pad inside of their mouths. They **grind** their food against this pad to turn it into a paste. A platypus may pick up **gravel** in their cheek pouches to help them do this!

Lots of Food

Since platypus meals are so small, they must spend nearly 12 hours every day hunting for food! A platypus can eat half its body weight in food each day. It's a good thing they have such cool receptors to help them find bugs!

GLOSSARY

coarse: Hair or fur that is thick and rough.

electricity: A kind of energy that flows and is made by the movements of animals.

gravel: Small, loose pieces of rock.

grind: To crush something to powder or paste.

mammal: A warm-blooded animal that has a backbone and hair, breathes air, and feeds milk to its young.

muscle: One of the parts of the body that allow movement.

pouch: A pocket of skin inside the mouth of an animal in which it carries food.

receptor: Part of an animal's body that senses changes in its surroundings, like hot or cold.

surface: The top of a liquid.

webbed: Connected by skin.

FOR MORE INFORMATION

BOOKS

Kenney, Karen Latchana. *Platypuses.* Minneapolis, MN: Bellwether Media, Inc., 2021.

Sexton, Colleen A. *Platypus.* Minneapolis, MN: Bearport Publishing Company, 2021.

WEBSITES

National Geographic Kids: Duck-Billed Platypus
www.kids.nationalgeographic.com/animals/mammals/facts/platypus
Check out maps, photos, and more facts about platypuses.

San Diego Zoo Wildlife Explorers: Platypus
www.sdzwildlifeexplorers.org/animals/platypus
Learn more about platypuses and how they survive in the wild.

Publisher's note to educators and parents: Our editors have carefully reviewed these websites to ensure that they are suitable for students. Many websites change frequently, however, and we cannot guarantee that a site's future contents will continue to meet our high standards of quality and educational value. Be advised that students should be closely supervised whenever they access the internet.

INDEX